Ergonomics in Everyday Life

Ergonomics is often misunderstood as a field reserved for experts in engineering, psychology or design, leaving many feeling disconnected from its relevance. In reality, ergonomics is deeply embedded in our daily lives, shaping everything from how we wake up to how we go to bed.

This book simplifies the complexities of ergonomics into relatable concepts, making it accessible to everyone, regardless of their background. Through a narrative approach enriched with practical examples, it explores the ergonomic aspects of daily routines, offering actionable advice to enhance comfort and efficiency. Each chapter begins with an overview to set the stage, concludes with key takeaways summarising the content and includes practice sessions to help readers apply what they've learned. The final chapter features exercises designed to reinforce ergonomic principles in real-life scenarios.

Written in a concise and engaging style, *Ergonomics in Everyday Life: Simplifying the Science of Comfort and Efficiency* empowers readers, whether professionals, engineers, designers or individuals with no prior knowledge, to understand and apply ergonomic principles to improve their daily lives.

Ergonomics in Everyday Life

Simplifying the Science of Comfort and Efficiency

Prabir Mukhopadhyay, MSc, PhD

CRC Press
Taylor & Francis Group
Boca Raton London New York

CRC Press is an imprint of the
Taylor & Francis Group, an **informa** business

Designed cover image: Shutterstock

First edition published 2027
by CRC Press
2385 NW Executive Center Drive, Suite 320, Boca Raton FL 33431

and by CRC Press
4 Park Square, Milton Park, Abingdon, Oxon, OX14 4RN

Press is an imprint of Taylor & Francis Group, LLC

ISBN: 9781041376316 (hbk)
ISBN: 9781041376347 (pbk)
ISBN: 9781042009152 (ebk)

DOI: 10.1201/9781042009152

Typeset in Adobe Caslon Pro
by KnowledgeWorks Global Ltd.

To the Lotus Feet of Shri Ramakrishna
Paramhansha Dev, Pronam to You.

To Late Dhirendranath Mukhopadhyay (my Father)
&
Mrs Meena Mukhopadhyay (Mother), Dada,
Mamoni, Duggi and Leto . . .
&
To the common person struggling hard to make both ends meet, with the hope that this book brings them some comfort in their daily struggle for survival.

Contents

About the Author			xi
Preface			xiii
Acknowledgement			xv
Chapter 1	Layperson: Myth or Reality?		1
	1.1	Overview	1
	1.2	Laypersons: Who Are They?	1
	1.3	What Do Laypersons Do?	2
	1.4	Significance of Laypersons in Society	4
	1.5	Ergonomics and Laypersons	4
	1.6	Daily Life and Ergonomics	4
	1.7	Relationships: People–People, People–Environment, People–Product	5
	1.8	The Pillar of Laypersons: Their Reliability	6
	1.9	Recap	6
	1.10	Practice Session	7
Chapter 2	The Dawn		9
	2.1	Overview	9
	2.2	Ringing of the Morning Alarm	9
	2.3	Getting Up and Down from the Bed	10
	2.4	Navigating to the Washroom in the Dark	11
	2.5	Exact Location for Urinating	12
	2.6	Switching on the Room Light: Location	12

2.7 Brushing the Teeth, Hand and Handle 14
2.8 Rinsing the Mouth, Sink Height, Tap Access, Mirror Height 15
2.9 Recap 16
2.10 Practice Session 16

CHAPTER 3 DAY BEGINS 19
3.1 Overview 19
3.2 Morning Cup of Tea 19
3.3 Breakfast Table 20
3.4 Eye on Time: The Wall Clock 23
3.5 People for Breakfast: Too Many 24
3.6 Communication while Eating 24
3.7 Jam Bottle, Plate Dimension, Knife and Fork 25
3.8 Hand Towel Customisation 26
3.9 Recap 27
3.10 Practice Session 27

CHAPTER 4 TRAVEL TO WORK 29
4.1 Overview 29
4.2 Wardrobe Categorisation 29
4.3 Tying the Knot 31
4.4 Bathroom Shower 32
4.5 Hair Combing and the Mirror 32
4.6 Personal Grooming 33
4.7 Shoe Polishing 34
4.8 Shirt, Trousers and Good Fit and Travel to Work 35
4.9 Recap 36
4.10 Practice Session 37

CHAPTER 5 WORK AND ONLY WORK 39
5.1 Overview 39
5.2 My Workplace 39
5.3 Dichotomy in Space 41
5.4 Colour and Music at Work 41
5.5 Body Clock and Efficiency at 4 pm 44
5.6 Lunchtime 45
5.7 Claustrophobia in Elevator 46
5.8 The Torture Chamber of the Boss 47
5.9 Recap 48
5.10 Practice Session 48

CHAPTER 6 BACK HOME 51
6.1 Overview 51
6.2 Friday: The Weekend Joy 51
6.3 Crowd on Road and Control 52
6.4 Same Place but Different at Night 54

6.5 Along the Known Path and the Signages 54
6.6 Freshen Up and the Confined Space 55
6.7 Different Clothes, Different Moods 55
6.8 Dinner and the Television 56
6.9 Recap 56
6.10 Practice Session 56

CHAPTER 7 DUSK 59
7.1 Overview 59
7.2 The Nightdress for Sleeping 59
7.3 Planning for the Next Day 60
7.4 Sleep and the Calmness of the Room 61
7.5 Switching Off Digital Screens 61
7.6 Water and Medicine at Bedside, Reminder 62
7.7 In Bed, Dark and My Position 63
7.8 Dream Sweet Dream 63
7.9 Recap 64
7.10 Practice Session 64

CHAPTER 8 ACTIVITIES WITH ERGONOMIC DIRECTIONS 67
8.1 Overview 67

BIBLIOGRAPHY 75

About the Author

Prabir Mukhopadhyay is an Associate Professor in the Design Discipline at the Indian Institute of Information Technology, Design and Manufacturing (IIITDM), Jabalpur. He earned his PhD in Industrial Ergonomics from the University of Limerick, Ireland. His research interests encompass diverse applications of ergonomics, including manufacturing, the informal sector and product, space and communication ergonomics. Dr Mukhopadhyay has collaborated on numerous consultancy projects in ergonomics, working with esteemed clients such as Indian Railways, the Self-Employed Women's Association (SEWA), the United Nations Industrial Development Organization (UNIDO) and Mazagon Dock Shipbuilders Limited. He is the author of six books published by CRC Press and enjoys writing, travelling and cooking in his leisure time.

Preface

The subject ergonomics is known by very few in society. People confuse it with different subjects like economics, agronomy and so on. For laypersons, it's a subject which seems to be alien and of no use to them. If this is the attitude of the layperson towards a subject, then that subject can never grow and become popular. To make a subject popular among the common people, it's necessary that the subject be narrated to them in a manner that they are able to relate to them.

This book makes an attempt in that direction. It tries to depict the subject ergonomics through the common person and how it is applicable in all spheres of her/his daily life, starting with getting up from the bed in the morning to going to bed at night. The writing is in a storytelling format, with examples from the real world, which should make the book interesting to read and at the same time easy to apply in the real world. Each chapter touches upon different aspects of the life of a common person and her/his struggle for survival. It depicts in a clear, storytelling format the exact connection of ergonomics at each and every stage of the person's life. The chapters start with an "overview" section which prepares the readers for what they can expect out of the chapter and ends with "Recap", which is the gist of the chapter. This section is followed by a practice session with ergonomic directions which would help the readers test how much they have understood. The last chapter is dedicated to different types of exercises with ergonomic directions. The exercises have been carefully selected from the daily life of a common person and the different ergonomic applications. All characters named in this book are fictitious and do not have any resemblance to anyone living or dead. Any resemblance if found is purely coincidental.

I hope the readers enjoy reading the book. Any comments and suggestions are always welcome.

Dr Prabir Mukhopadhyay
Jabalpur

Acknowledgement

All the diagrams in this book have been created by two of my Bachelor of Design students, Akhil Negi and Kritika Kakkar. Thank you so much Akhil and Kritika for breathing life into the book!

1

Layperson

Myth or Reality?

1.1 Overview

This chapter provides the readers an overview of the subject ergonomics from the perspective of a layperson. The term *layperson* is often misunderstood and does not mean that the person "does not know anything". On the contrary, the term "layperson" is a misnomer as humans are the most evolved species on the globe. These people are already using ergonomics in their daily lives without even knowing it. The chapter reveals how ergonomics features in daily life and how important it is for our survival and well-being. It helps make different relationships between people, between people and tasks, between people and the objects around them stronger and ensures the seamless integration of humans with everything around them. This makes the life of the common person beautiful and fruitful.

1.2 Laypersons: Who Are They?

Laypersons are all around us. We are all laypersons when it comes to specific fields, and some of us are experts in certain fields (Figure 1.1). The doctor is an expert in medical sciences or disease but might not be an expert in bridge construction. Thus, laypersons are the people who comprise society, regions and countries. They have some aims and objectives in life, of which the two prime ones are possibly health and happiness. In this book, these are the people who are being referred to. The ones you meet on the road, see on public transport or watch soccer matches with. Some of them are friends, some acquaintances and some

DOI: 10.1201/9781042009152-1

Figure 1.1 The layperson.

strangers. Each has their own struggle in life, their problems, their likings and their way of looking at life. Though there are differences, there are some common elements which every layperson has to face. Right from getting up in the morning and going to work, or study, and returning home in the evening and going to bed, they get in touch with other different laypersons (we call them people), places, products, services, etc., which affect their lives in many different ways. Thus, laypersons are you and I and are not a myth.

1.3 What Do Laypersons Do?

A layperson is the pillar of this world (Figure 1.2). The world is a better place because of laypersons. On the other hand, the world

Figure 1.2 Laypersons as society builders.

is a dangerous place at times because of them. They work hard to earn a living, and so do their families. They expect that the work they do is a little more humane and enjoyable. These are the people interacting with others, space and objects and trying to reach a target or achieve their specific goals in life. At times, the goals are doing their job efficiently at their workplace, sometimes achieving their goals at the family front, educating children, maintaining the family and hosting various activities. When these laypersons are able to do their job properly, we appreciate them, but the moment they make mistakes, which might be a small one at the workplace or a mistake made in the kitchen which spoils the food, we blame them and label them "unreliable".

1.4 Significance of Laypersons in Society

These are those who actually are the building blocks of society and any system around. They hold society and the world, and that's why the world is a beautiful place to live and enjoy as long as we are here. They contribute; they produce; they are the building blocks holding society together. They work hard, hand in hand with others; they make objects (we call products); they augment their capability using tools and machines. They work in a context or space; they communicate with one another; and ultimately they do all these to sustain their eternal struggle for existence. Thus, without them society doesn't exist; they are the society, not mere "pillars" of the same.

1.5 Ergonomics and Laypersons

Ergonomics means the interplay between people, objects and the environment for enhancing productivity or achieving a common goal. A young lady trying to pluck ripe mangos from the trees is unable to do so as she is too short. She gets a big pole (a mango plucker) and while remaining on the ground is able to pluck the mango with ease. The pole actually amplifies the capability of the lady's arms or becomes an extension of her arms and helps her achieve what she wants (Figure 1.3). This is exactly what ergonomics tries to do: amplify human capability based upon people's needs. Since people are an integral part of the subject of ergonomics, it's all about working towards the benefit of people. From a pin to an aeroplane, ergonomics looks to ensure that they benefit the layperson in all possible ways. In this world, everything that is done is ultimately for the benefit of humankind. This is where ergonomics and the layperson come closer more and more.

1.6 Daily Life and Ergonomics

There is a bit of ergonomics in our lives. All around, you see it about the layperson and their interaction with people, space, objects, services, etc. The moment you get up from bed, the dimension of the

Figure 1.3 Mango ready to harvest.

bed has to be in accordance with your body dimension, and the room temperature has to be comfortable or else you won't have a good sleep. You should be able to set the alarm correctly, or else it won't wake you up at the right time. We are going to narrate in this book how your daily life is influenced by ergonomics.

1.7 Relationships: People–People, People–Environment, People–Product

Relationship is the key to ensuring a better society and efficiency. Though each person is different in all ways, they have their similarities. There are people-to-people relationships as colleagues and friends; people-to-environment relationships which are the context in which people work, travel and rest; and people-and-product relationships. These relationships are the bridges that close the gap between people, objects and space and at the same time ensure that they are seamless and fully integrated. Relationship is the basis on which we all exist. Ergonomics as a subject acts as a binding force in keeping these relationships intact and takes care that they do not split apart.

1.8 The Pillar of Laypersons: Their Reliability

People are the most reliable species in this world (Figure 1.4). I reiterate what I said at the beginning that we are the most evolved species on this earth. When systems fail, it's not wise to blame people. Systems fail not because people made mistakes, but because we have not been able to address people's needs and wants. We have not accounted for people's capacity and limitations and have not taken countermeasures for the same. Let's believe that the layperson deliberately never makes a mistake. She/he also is competing with other species to survive in this world. This competition is needed to ensure that they always remain at the top and are not overtaken by other species.

1.9 Recap

a. Laypersons comprise the society.
b. Ergonomics makes layperson life humane.

Figure 1.4 People are the most reliable species on earth.

c. Laypersons are the pillars of the society.
d. As the most evolved species, laypersons cannot make mistakes.
e. Mistakes happen when people are not accounted for in any people, product or environmental system.
f. The world revolves around the layperson, mainly their health and happiness.
g. Ergonomics, the science for people, is all about the layperson.

1.10 Practice Session

a. Look around as you go to the market for buying your daily groceries. List the different types of people that you see in the market who visit as customers. Try to observe how differently they interact with the shopkeepers while buying.

Directions:

i. Observation means using all your senses to analyse what is happening in front of you.
ii. Question yourself as to why a particular interaction takes place in a certain way.
iii. Try to find out the different conversations between the people and the shopkeepers.
iv. Try listing the ergonomic issues.

b. Visit your nearest post office. Try to identify the different ergonomic connections that the common people who come to that place have to face. Make sure you list those issues which are relevant for getting a better service at the post office.

Directions:

i. Try listing different touch points for people and the post office.
ii. See if there are any relationships between the different touch points and the people.

iii. List those places where ergonomic intervention can happen.
iv. Do you see any good ergonomics in the place from the viewpoint of the laypersons?

c. Look at your house and try to locate a place where there is maximum interaction with outsiders and the members of your house. List the different ways in which this interaction could be improved from an ergonomic perspective.

Directions:

i. List the place where maximum people from outside sit to interact with your family members.
ii. Try to see if the space allows for the movement of all people in the space.
iii. Is there adequate light and ventilation in the space?
iv. What about the furniture? Are the furnishings comfortable?
v. What do guests say (good) about the house when they leave?

2
The Dawn

2.1 Overview

The chapter begins with the life of a common person as the day breaks. It narrates the different ergonomic issues which are associated with all bits and pieces of our daily life, of which we are not aware as such and ignore. Right from the ringing of the alarm bell to going to the washroom for brushing our teeth, the value of ergonomics is depicted everywhere. The chapter also takes the readers through the different objects at home like the chair, table, sink and tap, which are essential for any home, but their importance from an ergonomic perspective is narrated. The chapter takes the readers through the different daily tasks with their ergonomic connections.

2.2 Ringing of the Morning Alarm

It is 4.30 am and the alarm bell keeps ringing. Mami and Nitai are fast asleep. It's still very cold and dark outside, but this couple have to get up and leave for their respective workplaces far off from their home. Mami asks Nitai, "please shut off the alarm". Nitai says, "you do it, I cannot see it properly from the bed". The room is dark, and as the couple were sleeping, they were unable to see the location of the alarm clock in the dark. It takes time for the eyes to adapt to the darkness, a phenomenon called "dark adaptation". For Mami, the alarm clock dials are not "legible", that is, she cannot understand the clock hand positions because of her distance. Possibly while designing the dial of the alarm clock, one should factor in this aspect as to the distance from which people need to see the clock hand positions. If the alarm can be easily turned off by a mere tap from the bed, a person might go to sleep once again. So, from the layperson's perspective, it should be difficult to switch off the alarm, and

DOI: 10.1201/9781042009152-2

Figure 2.1 The alarm clock and its switching off mechanism.

the level of difficulty should be such that you need to get out of the bed, manipulate the alarm and then switch it off. By that time, you are already awake because of the activity you just performed (Figure 2.1).

2.3 Getting Up and Down from the Bed

Mami was about to get down from the bed, and her legs from the surface of the bed were dangling. That means she would have the same problem while trying to get on the bed. Here also the ergonomics of human body dimensions need to be factored in. The height of the bed should be calculated such that while Mami sits on the bed, her legs should comfortably touch the ground. This would

Figure 2.2 Problems of Nitai and Mami sleeping on the bed.

also ensure that she is able to get up on the bed with ease. Nitai is taller than Mami and doesn't have any issues in getting up and down off the bed. His problems are elsewhere. The length of the bed is too short for Nitai and it touches his feet. Nitai has to fold his legs while sleeping. Thus, while determining the length of the bed, one should factor in the height of the person plus consider the fact that a person while sleeping moves up and down! Even Mami complains that Nitai while sleeping rolls from one end of the bed to the other and at times pushing Mami to the edge of the bed! Here again, for determining the width of the bed, one needs to consider not only the body width of two people but also, add to all this, the different movements exhibited by people sleeping on the bed (Figure 2.2).

2.4 Navigating to the Washroom in the Dark

Nitai goes to the bathroom, and Mami sits on the bed. Interestingly, Nitai can navigate in the dark to the bathroom, locate the toilet and urinate. He takes the toothbrush in the dark and brushes his teeth all in the dark without switching on the lights, because Mami is still in bed and gets irritated at the lights. How is this possible? Another miracle of ergonomics is that people develop a typical navigational pattern in their own homes where they live for a long time. With the help of this conditioning, they are able to navigate accurately in the dark at their familiar places like their home. This is not possible when you check into a hotel. There you keep the bathroom lights on so that you can locate the bathroom if you have to use it at night.

2.5 Exact Location for Urinating

Nitai has the capability of urinating in the commode in the dark and can aim very accurately. This is also possible through conditioning in his familiar place without the need to switch on the lights or touching anything around to locate the commode. Here the commode height is important because it should ensure that the person sitting on it, like a chair, should be able to rest his/her feet firmly on the ground and should not dangle. Thus, it's better to consider the one with shorter leg length. The diameter of the commode hole should be designed considering the buttock diameter of the shorter person. If it's done taking the dimension of the larger buttocks, then the person with smaller buttocks might go inside the commode! The handle of the commode should be within easy reach of the shorter person, and the handle should be easy to operate with least force. This is to ensure that women and elderly people are able to operate them with maximum ease (Figure 2.3). Nitai on many occasions when the bathroom was newly constructed urinated just beside the commode instead of inside it. The reason possibly was that he was not used to the space, or the space was not mapped clearly in his mind until later.

2.6 Switching on the Room Light: Location

The switchboard for lights should be in the place where you expect it to be. If the switchboard is behind a door, it becomes difficult to

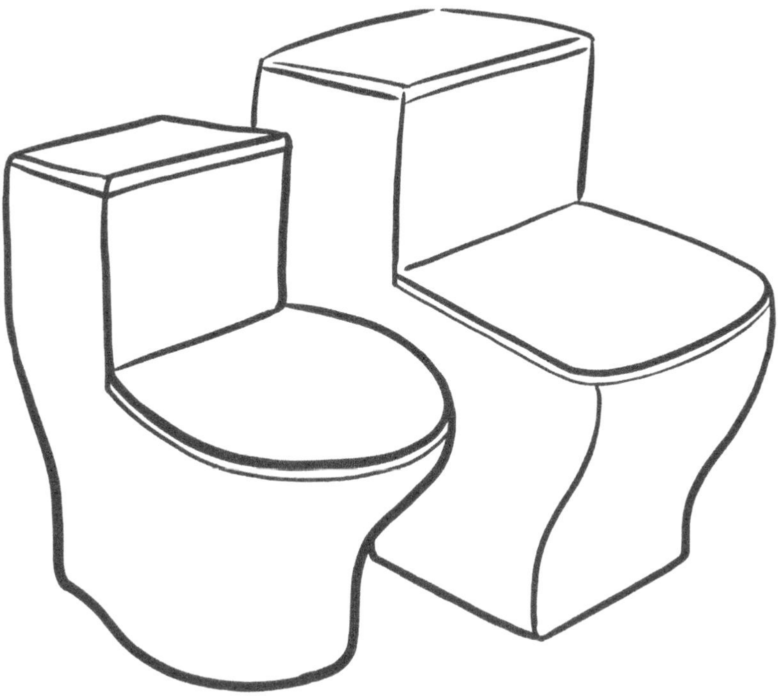

Figure 2.3 Commode design with respect to the users.

locate it and not be able to access it easily. Secondly, switch boards should be in proximity to the lights and fans they are connected with so that people are able to relate to them easily. On the board, the switches should be differentiated either by their position, texture or colour. This is to ensure that the switches do not confuse the user. Mami always confuses the switches for the light and fan. Thus, in winter when she wants to switch on the lights, she by mistake switches on the fan, and this irritates Nitai who keeps on shouting. Nitai has coloured the light switches in pink and the fan switches in yellow, so that Mami does not confuse them. The reason is that the fan is yellow coloured and the lamp shades are pink coloured. So the switches are now coloured to indicate their "relationship" with the respective devices. Moreover, these two colours are also visible in low-light conditions. Interestingly, a week before, Mami's elderly parents came to stay with them for a week. It was extremely difficult for them to locate the switchboard at night

Figure 2.4 Locating the switchboard and the switches in the dark.

and to identify the specific switches. They were switching on all the lights and fans in the middle of the night and thus waking up everyone. It took some three to four days for them to get adjusted, and by the time they were adjusted to the switch boards' position and layout, it was time for them to leave the place (Figure 2.4).

2.7 Brushing the Teeth, Hand and Handle

Sukumar, the son of Nitai and Mami, is just back home for his annual leave from Eldorado. He has been in Eldorado for 15 years and is working there as a scientist. Sukumar goes to the washroom in the morning to find a new set of toothbrush and toothpaste for him. He takes the toothbrush in hand and finds the handle too thin and the length of the brush too small to hold properly. When his hand was wet, the brush almost slipped from his hand, and it was

Figure 2.5 Mismatch of the toothbrush and the hand while brushing the teeth in the morning.

difficult for him to use the toothbrush in cleaning his teeth. Again, a case of ergonomic mismatch. The toothbrush should be designed in tandem with the hand dimensions, and if it's too small (in this case) or too large, manipulating it would be difficult. It's supposed to be an extension of the hand for cleaning your teeth. You need to factor in the context of use of the toothbrush, and in this case it's the hand full of water and the liquid toothpaste dripping down the brush handle to the hand (Figure 2.5).

2.8 Rinsing the Mouth, Sink Height, Tap Access, Mirror Height

Sukumar's son (Jogai) enters the washroom next. He is only eight years old, and this is his first visit to this house after he was born. Jogai wanted to wash his face, but the sink height was near his chin. He tried stretching his hand to reach for the tap, but unfortunately it was far beyond his reach. He found the mirror in the washroom too high and could not see his face. There were so many ergonomic mismatches, and it seemed that the washroom was designed only for adult usage and not for children like Jogai. Jogai thus leaves the washroom in frustration without brushing his teeth and washing his face. There are so many mismatches between the sink, tap and mirror. So while designing such specific spaces in the house, you need to keep in mind the people of different age groups ranging from children to the elderly who need to be accommodated.

2.9 Recap

i. An alarm clock and other allied important elements for waking up should be within easy reach and easy to shut off with one operation.
ii. The bed dimension should factor in people of different dimensions: tall and short, fat and thin, young, elderly and children.
iii. The navigation route to the washroom at night is difficult for those who are new to the place and could be augmented by either night lamps or floor lighting.
iv. Washroom design should factor in different contexts of use.

2.10 Practice Session

a. You need to get up early to catch the early morning flight. You are alone at home and need to ensure that you wake up on time. How would you ensure this from an ergonomic perspective?

Directions:

i. Use redundancies. That means use the alarm in your clock or mobile phone and ask the driver of your car to come early and wake you up in case you are not awake.
ii. Do not pull the curtain in your room so that as the outside becomes lighter, you get feedback.
iii. Make sure you go to bed a little early so that you automatically get up early.

b. You have checked into a hotel for the night. You find that the room is too big and that navigating to the washroom would be difficult at night with all the lights off. From an ergonomic perspective, how do you address this problem?

Directions:

i. Keep the washroom lights on at night.
ii. Sleep on that side of the bed which opens towards the bathroom.
iii. If there are floor lights, keep them on.
iv. Make sure there are no obstacles on the path to the bathroom.

c. You have gone to your uncle's home and have decided to spend the night there. You find that the switchboard in the living room is not in an accessible location and the layout of the switches is such that it's difficult to identify the lights and fans. How can you help your uncle address this ergonomic issue?

Directions:

i. Paint the switchboard in a fluorescent colour to enhance visibility from a distance.
ii. Give a thick border around the board.
iii. Label the switches for lights, fans, etc.
iv. Colour-code each light and fan and use the same colours on the respective switches.

3
Day Begins

3.1 Overview

This chapter is an overview of the different ergonomic connections in the daily life of any layperson, starting with the morning cup of tea, its preparation, to the breakfast table and dining associated with different users and their problems through the lens of ergonomics. Simple table gestures like communication and how they facilitate interactions are discussed. Small elements at the breakfast table and their ergonomic issues related to their usage are all discussed here through different characters which represent the layperson. The chapter in brief gives the readers a snapshot of different ergonomic issues in what we do daily without ever realising the same.

3.2 Morning Cup of Tea

Priyanshu is just back to India from the United States with his parents. The young man of 22 years old has a habit of drinking green tea in the morning sharp at 5:30 am. Unfortunately, his parents are still asleep, and this is his first morning since coming home after two years. He starts looking for the electric kettle he is used to in the United States for making tea. He is unable to locate it. Finally, he ends up with a saucepan. He then looks around for that cup, sieve and more importantly the green tea. Priyanshu is used to the kitchen at his home, and things are organised there in a particular manner. So he has in his mind the same image he is used to. There is a fanciful name for this "mental model" in ergonomics. He manages to boil the water and put in some green tea leaves. After the leaves are soaked for some time, he gets hold of the teacup and takes it to the verandah to sit and enjoy his cup of tea. He notices that the handle of the cup is too small for him to hold with all his fingers, and he

DOI: 10.1201/9781042009152-3

can insert only two fingers and use the thumb for directing the cup towards his lips. Again, there was a mismatch between his hands and the handle. Thus, the design of the cup handle should consider that people are "different"; some have a larger palm and some smaller. Ergonomics suggests that one should be a little biased towards the larger palm, because in that case the smaller palm would fit easily. Over and above in extreme cold weather, people might be wearing gloves, and the hand should be able to accommodate the gloved hand and the extra space for the same. There could be another problem: if the hands are moist, then holding the cups will be difficult. So, the context of product use is also important, not the person holding it alone. "So much ergonomics in just drinking tea! Who knew this?" Priyanshu thought as he sat down to sip his cup of tea while looking at the mountain ranges. So many ergonomic issues, the cup design, locating different ingredients and expecting them to be in the same sequence as his home, etc., "The mesmerising world of ergonomics!" Priyanshu thought.

3.3 Breakfast Table

Mami gets up at eight o'clock and finds Priyanshu on the balcony sipping his cup of tea. "Hello, Priyanshu, so did you manage to make your cup of tea?" she asked. "Yes, with a lot of struggle", Priyanshu replies. Mami goes to the kitchen to cook breakfast for all. In the kitchen, she is having a problem with the height of the kitchen slab. It seems to be too low, and she is having low back pain. The reason is that the kitchen platform for Mami is a place where she does multiple tasks associated with cooking. She washes, chops vegetables and cooks as well. So ideally the platform should be close to her elbow height or a little below. In the current scenario, since she is of late using high-heeled slippers, her height has increased, and the kitchen platform now has gone down near her hip. This is forcing her to stoop and thus leading to back pain. One solution is for her to use a slipper with a very thin sole; another is for her to do most of the job on the platform on another wooden plank. Again ergonomics! Breaking and reconstructing the kitchen platform is an expensive affair, and few would opt for it.

Mami cooks some fried eggs, bakes some toast and chops a few salads. She gets some fruits like grapes and pears and moves towards the breakfast table. She calls Mama and Priyanshu for breakfast. While sitting down at the table, Priyanshu notices that his elbows are touching that of Mama's, who is sitting at the end of the table, and even his plate is also in close contact with that of Mami's plate sitting opposite him. This is a typical problem while sitting at any table for different purposes. Every person needs a personal space of her/his own, where no one should intrude. This personal space can be mapped by roughly drawing a circle with the head at the centre and the two elbows spread on the two sides of the body (Figure 3.1) as the diameter. The table should give space to each individual, and in this zone the person's plate should be positioned. The common food items like jam, butter and fruit bowls should be located outside the personal zone so that everyone can reach and access them. Mami was trying to open the new jam bottle. Mama keeps smiling. Upon carefully looking at the jam bottle, Priyanshu finds that it requires grip and torque in the anticlockwise direction. For females, this is difficult because of their forearm muscle

Figure 3.1 Priyanshu at the breakfast table.

Figure 3.2 Mama is unable to open the moist water bottle.

structure. So Mami reaches out to Mama who opens it at the flick of a second. Mami takes out the water bottle from the fridge and keeps it on the table. Everyone is enjoying their breakfast. Mama reaches for the water bottle but is unable to open it (Figure 3.2). Priyanshu notices that the bottle is moist due to condensation as it was in the fridge. This was leading to slippage of the palm while trying to open the bottle. The context of use is very important, and those who designed the bottle just imagined that the bottle would be used in dry conditions and not wet. In these cases, texturing on the bottle surface helps as it increases friction.

Mami was trying to cut the omelette on her plate into smaller pieces, but there was another problem. The handle of the fork is too small, and hence Mami did not have a good grip on it. So, the handle is the extension of the hand and should fit the profile of the hand and thus needs to be designed accordingly. Priyanshu receives a phone call from his office while at the breakfast table. He turns his back towards the table and starts communicating. It looks odd.

At the table while dining people need to interact, and for that they need to sit face to face with eye contact. The moment eye contact is gone, you become a stranger! This type of seating that is back towards the person is preferred in the public domain where you encounter a stranger. But when you are with your dear ones, it's not a preferred mode of seating; here you prefer eye contact for facilitating interaction. This is applicable in any group setting.

3.4 Eye on Time: The Wall Clock

While dining, Priyanshu was keeping an eye on the clock because he must reach his office on time. He finds that the clock is not within his visual cone, and there is a pillar in the room which obstructs the full view of the clock. So, while placing the clock in any room, again the context is important. One must think of how many ways and from where in the room people would need to see the clock and the time. Here comes the question of which clock is better: digital or analogue? It depends on what purpose the clock serves. If you want to know the accurate time, then digital is better. This is suitable for industrial settings where accuracy is very important. In a domestic setting, you need the clock to give you a feel of the time in terms of its "rate of change". That is, how long until 8 am is it? Still a few minutes to go or more? This fills in space for analogue clocks to score them above digital in domestic settings (Figure 3.3).

Figure 3.3 The analogue vs. digital clock debate.

3.5 People for Breakfast: Too Many

As Priyanshu was having his breakfast, the doorbell rang. He gets up and opens the main door to find four of Mama's friends who would like to join them for breakfast. The four people sit at the table only to find that they are too cramped and that there is not enough space at the table to accommodate four extra people. Each person's elbow touches one another's. When Mami kept their respective plates on the table, the plates were touching each other, thus showing that the personal space of each one had been "intruded", leading to discomfort at the breakfast table (Figure 3.4).

3.6 Communication while Eating

All seven people are now dining at the table and enjoying their meal. Though the space was cramped, they seemed to be enjoying the company of each other. They were facing each other, which facilitated communication. Just imagine that each person is made to sit inside a closed cabin next to each other and asked to dine! They can talk to each other but cannot see! Weird, isn't it?

Figure 3.4 Discomfort at the breakfast table due to too many people.

This is because eye-to-eye contact facilitates better interaction and puts us at ease. That is why while you dine out with your friends, you sit facing each other.

3.7 Jam Bottle, Plate Dimension, Knife and Fork

Bhola, one of the elderly people at the breakfast table, tried to pick up the jam bottle. Unfortunately, it's too thick for him to hold, and the bottle drops on the table spilling the jam all around. Priyanshu gets up from his chair and rushes to the rescue of Bhola who was very embarrassed. Upon close investigation, it was found that the diameter of the jam bottle was much larger than the palm length of Bhola. Thus, when Bhola tried lifting the jam bottle with his palm by curling the fingers around the bottle, his fingers did not overlap to ensure a proper grip. Thus, he lost control of the jam bottle, and it fell. Again, an example of mismatch in the bottle dimensions and the dimension of the palm which has not been considered. People are different in terms of everybody's dimensions, and this aspect needs to be considered in designing anything for daily use, including the jam bottle. By the time Bhola settled down at the breakfast table, Kashi, another morning walker, started complaining about the plate being too small. The problem Priyanshu noticed was that Kashi had filled up the plate with ten slices of toast, eight boiled eggs and two big pears (Figure 3.5)! It was obvious that there was hardly any space on the plate where he could grip the plate a little with his fingers as he supported the same on his palm. Thus, food was falling from his plate. This is a case of extremes! No one expects a person to have so much food for breakfast. Plates are designed keeping in mind that people would be generous while eating and not load their plates with food beyond a limit for average ones! So, it's anticipated that there would be some free space for gripping the plate with the fingers with the food on the plate. Unfortunately, extremes are a part of the population, and unless it's an emergency, we do not consider them. Otherwise, the production cost jacks up. If we must consider the case of Kashi, then one has to design a huge plate just for a few people like him!

Figure 3.5 Kashi trying to overload his plate.

While the above incidents were "entertaining" Priyanshu at the breakfast table, there was a problem with Sital, another morning walker. He was complaining about the knives and fork handles being too thick for him to hold. He was pointing at the wooden handle of the knife and the fork and saying that it's difficult to manipulate them. This was again a mismatch issue. The good part of the wooden handle is that it prevents slipping from the palm in case the fingers are moist while dining. So not everything is bad!

3.8 Hand Towel Customisation

After breakfast everyone got up, and some moved towards the washing basin to wash their hands. Priyanshu could locate his hand towel, which was yellow, and similarly Mami's towel was red, and Mama's towel was green. The colour helped people differentiate their towels. This is good practice from an ergonomic perspective when it comes to differentiating between similar-looking products, in this case hand towels. The guests were provided with disposable paper towels.

3.9 Recap

a. Products of daily use should fit the human body parts.
b. Factor in the context of use of any product while designing it.
c. Extremes of the population are a reality; we need to factor them in an emergency but not everywhere.
d. Colour is one of the many elements for differentiating similar-looking products.

3.10 Practice Session

a. Sitala is at the dining table in a small hotel. She finds the table to be extremely small for dining as she needs to keep her laptop open while eating. You must find an ergonomic approach for designing such a table for her.

   Directions:

   i. The table surface should factor in the different elements and its dimensions.
   ii. The movement of the user at the table surface should be considered.
   iii. Consider while designing the table surface how much the user can reach by stretching her arms and how much she can cover.
   iv. Keep some space for another person to sit on the opposite side.

b. A company has approached you to design their tea mug. You have to apply ergonomic principles in designing the same. How would you proceed?

   Directions:

   i. Identify the touch points of the palm and the parts of the mug.

ii. Think of the higher and the lower values.
iii. For reach/access, aim for the lower value.
iv. For clearance, aim for the higher value.
v. Factor in the context of use.

c. A newly opened restaurant seeks your advice as to whether seating should be arranged with people facing each other or should be arranged with users facing back-to-back.

Directions:

i. It should be a combination of both.
ii. Friends would prefer face-to-face seating.
iii. Strangers prefer back-to-back seating.
iv. Have the majority of the seating face-to-face.
v. A few of the seatings could be back-to-back for strangers.

4
Travel to Work

4.1 Overview

This chapter provides the readers an overview of the ergonomic principles involved in the daily life of a layperson as she/he prepares to leave for work. The ergonomic issues involved in picking up the right clothes, dressing up and personal grooming are all discussed in this chapter. The readers are introduced to the ergonomic principles which dictate a good fit of the clothes that we wear to work so that one looks smart and formal. Even small things like polishing one's shoes and the ergonomic principles involved are discussed here. The chapter gives the readers an insight into how ergonomic principles could be used in different areas while one travels to work from home and back and thus make one's life much more comfortable.

4.2 Wardrobe Categorisation

Vitto works as a software engineer at the IT Park in New Town. He has to get up in the morning and get ready to travel to work by public transport, which is the underground tube. The mornings are extremely hectic for him to prepare and get ready for travel to work. The choice where he gets confused every day is what to wear to work today. He stands in front of his wardrobe and gets confused (Figure 4.1). He starts at the shirts, trousers and ties and the different varieties and gets puzzled. Bapto, his younger brother, comes in and offers to help him out of this complex situation. "Look, Bro, you need to decide what you want to wear for five days of the week. Either you categorise your wear based upon shades, or you categorise them on the basis of any theme, which could be, say, Monday first day of the week, a start with blue . . . blue sky. Tuesday, your meeting day, switch to white and black trousers. Likewise, you plan for other days

DOI: 10.1201/9781042009152-4

Figure 4.1 The wardrobe confusion.

and arrange your clothes accordingly. Or as I said before", Bapto continues, "have maroon for Monday, turquoise for Tuesday and so on, I mean shades so it's easy to remember and make a call. This is using ergonomic principles of 'categorisation' for organising your wardrobe on the basis of a theme. This way, if you arrange your clothes, you will spend less time dressing; it also becomes easy for you to select and make a decision that you would like to dress like Friday on Monday, in case you wish to go for a change in the sequence." "Thank you, Bapto", Vitto says. Vitto plans to organise his clothes once he is back from the office this evening. He promises to do this as he wants to reduce his wardrobe time in the morning so that he can dedicate some of his time for his exercise in the morning.

4.3 Tying the Knot

Tying the tie knot is a herculean task for Vitto every morning before going to the office. He has practised it so many times, but still when in a hurry he forgets the sequence of steps and messes up the whole thing. You cannot blame him alone (Figure 4.2). Many of us tend to forget simple things, and in fact when we are in a hurry or tensed, we forget more. Our memory is of two types, short- and long-term memory. In the long-term memory (working memory), information is stored for a short period. Unfortunately, our short-term memory capacity is very low and hence it can't hold a lot of information, and thus we tend to forget. So, what's the way out? Either write it down, or present information in small chunks and in stages so that we do not forget. In case of tying the knot, Vitto can have a small video depicting the steps or have a diagrammatic representation of the different steps needed for tying a tie knot. This way handholding

Figure 4.2 Confusion with the tie knot.

for Vitto could be done and he would be able to complete the task without any hiccups and not be late getting to his office!

4.4 Bathroom Shower

Vitto is tall compared to other family members. He has problems when entering the house, as he bumps his head against the door frame. The biggest problem he faces is when he enters the bathroom for a shower. The shower faucet actually touches his head. Especially when he has soap all over his body and his face, he is unable to see. So, when he tries to position himself under the shower faucet by mere feel of the water, his head hits against the shower faucet and he is in pain. So, the height of the shower faucet should factor in tall people like Vitto and also consider the fact that with soap all over the face, a person is almost blind. A tall person is most likely to hit his head against the shower faucet and injure himself. Thus, while building a house and its rooms, please consider enough clearance for the tallest person. At the same time, one should also remember that the shortest person, perhaps your mother, can reach the kitchen platform; otherwise, who is going to cook food for you? (see Figure 4.3).

4.5 Hair Combing and the Mirror

When Vitto stands in front of the mirror, he confuses his right and left sides, as in the mirror whatever you see is laterally inverted. Most of the time, he treats his right hand as his left hand by looking at his image in the mirror and perceiving it as another person facing him. This confuses him further and he is late for the office, because he is unable to comb his hair properly, especially when he has to create a hairline on the left side of his head, and ends up creating it on the right side. This is one of the common problems people face when they have to refer to their right and left, and their images right and left or the right and left of a person in front of them. It requires a little bit of decoding on the part of the person. The person has to "imagine" herself/himself to be

Figure 4.3 The bathroom faucet hitting the head.

standing inside the mirror (Figure 4.4) and facing towards him where he is now standing. This has to be done fast; otherwise, it would take a lot of time. Practice and repeated practice help people in doing this fast.

4.6 Personal Grooming

Shaving his beard is a tedious task for Vitto, especially when it's cold in the morning. He has noticed, while shaving, the handle of the razor often slips from his hand. The reason is that the handle is wet and has soap on it. With soap and water on the handle, the hand is unable to grip it properly. This is again when the context of the use of a product has to be thought of. Depending upon such a context, the hand has to be textured, given a form which is easier to hold and manipulate. Shaving the beard requires fine manipulation by the thumb and the forefingers of the hand. So, a place for resting the thumb on the handle would be beneficial for shaving.

Figure 4.4 The hair combing paradox.

The blades have to adjust to the contour of the face, because our faces are not flat but curved at many places.

4.7 Shoe Polishing

Shoe polishing for many of us is tiring, especially when it has to be done using hand. We do it after wearing the shoe or at times even before wearing it. The handle of the brush has to fit the contour of the palm. It is better if there is a cloth piece through which the palm could be slipped and the handle held tight. For people who want to polish the shoe after wearing it, it becomes difficult. The first difficulty is in bending forward and polishing it, so that one gets a

Figure 4.5 Shoe brush handle.

better view of the front of the shoe. To prevent bending forward, a shoe brush with a long handle could be considered. The handle should be designed so that it fits the palm and the person gets a good command over the handle for brushing (Figure 4.5).

4.8 Shirt, Trousers and Good Fit and Travel to Work

Of late, Vitto has developed a big belly. He is having problems with his existing trousers and shirts. The trousers he is just able to manage by wearing them below the belly button. The shirts are creating a serious problem for him. While wearing the shirts, he finds that because of the belly size, he is unable to tug the shirt in front. For the trousers, he is finding a problem that while he sits, the trousers' length shrinks and exposes his entire shin bone and almost touches his knee joint. I believe many of us can relate to this. This is again a problem with fit or human body dimension and the clothing that we wear. First of all, while designing any clothing, one has to factor in the relevant body dimensions, and that's what your tailor does. Next what needs to be done is to consider the movement or the type of movements the person would exhibit. For example, Vitto's

Figure 4.6 Design of the shirt with a big belly.

big belly could be addressed easily if the length of the shirt were a little more in front compared to the back (Figure 4.6). Similarly, the problem with his trousers can be addressed if while deciding the length of the trousers, Vitto is made to sit and the length is calculated (Figure 4.7), and taking the dimension of the same while making him stand as well would give insight into the movements and thus help design the clothing accordingly, ensuring a better fit. These dimensional aspects in ergonomics are known as *anthropometry*.

4.9 Recap

a. While in a hurry, people are stressed, so categorising items is essential so that one may find them easily.
b. Our short-term memory capacity is limited. So, one needs to write down things to be done.

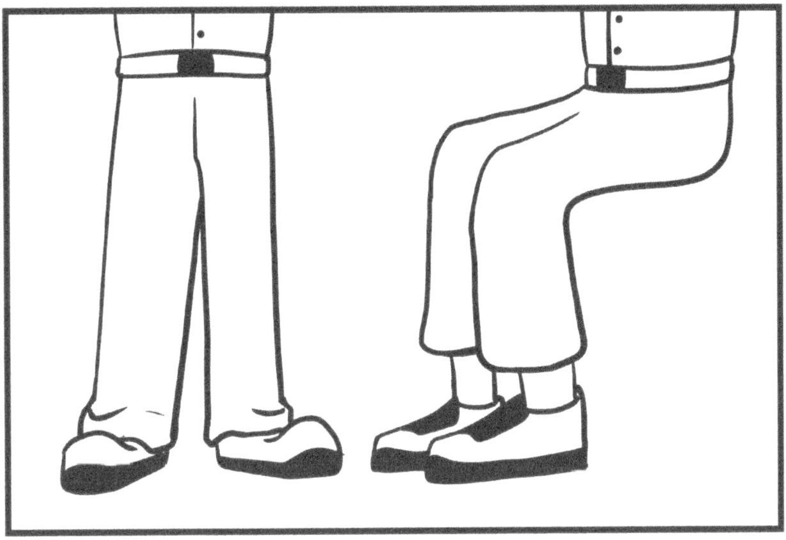

Figure 4.7 Design issue of the trousers while sitting and standing.

c. Visual representation of doing certain things helps, as people may not remember them.
d. Clothing design should account for different body movements to ensure a good fit.

4.10 Practice Session

a. Mr Kedar has twelve pairs of shoes. While going to the office in a different pair of shoes every day, he is unable to find out and decide which one to wear. How would you use ergonomic principles to help him out?

   Directions:

   i. Shoes are to be arranged serially as per days of the week.
   ii. Separate categories could be made for "office wear" and "casual wear".
   iii. Shoes could be kept in coloured bags.

b. While travelling to work, Ms Kalu forgets her umbrella every day in the tube coach. Can you find a solution for her?

Directions:

i. The umbrella should be connected to the purse with a long string.
ii. She could think of attaching the folded umbrella to her waist with a belt and making it a style statement.
iii. Hanging the folding umbrella on her neck.
iv. If she carries a bigger umbrella, then her purse should be integrated with it.

c. Ms Sundori is in a hurry every morning as she has to reach her workplace on time. The biggest hassle for her is grooming her hair. She has long hair down to her shoulder and does not wish to cut it. Your advice as an ergonomist is solicited.

Directions:

i. Inclusion of two mirrors, one in front and one behind, in her room.
ii. Keep all the elements for hair grooming ready on the dressing table the previous night.
iii. Arrange the oil, cream, comb and the clip in sequence the previous night.

5
Work and Only Work

5.1 Overview

This chapter provides the readers an overview of the ergonomic issues when the common person goes to work every day to earn her/his living. We see these problems every day but are not aware of them as ergonomic issues and how they impact our lives. The readers are taken through ergonomic principles in the workplace, including workplace monotony and how to deal with it. The confusion that a common person has in the workplace as to privacy and communication is discussed. The chapter talks about efficiency and how it keeps changing throughout the day and its ergonomic angle. The role of colour and music at work to make the work more enjoyable is discussed. How lunchtime brings together people and why people feel claustrophobic in the elevator are discussed. Lastly, how the office of the boss is perceived by the other staff in the office and the ergonomic issues associated with it are discussed.

5.2 My Workplace

Leto travels to his workplace, which is a biscuit-manufacturing unit. He is the supervisor of the packaging division and just like other supervisors has been given an individual office of his own. Leto feels that his workplace, both his private office and the assembly line which he is supposed to supervise, is just too boring and monotonous for him. This is common for any person. If they are exposed to the same place day in and out, it becomes boring and monotonous for them, and it's at that time they start "hating" their job. Ergonomics talks about making the workplace interesting! The question is, how? If we take the office of Leto, then there are two aspects of the office: one is comfort, and the second is "variety",

DOI: 10.1201/9781042009152-5

Figure 5.1 Making the workplace interesting.

which can add "flavour" to the office and make it interesting (Figure 5.1). For ensuring workplace comfort, the tables and chairs should be in tandem with Leto's body dimensions. For example, while seated, the table height should be close to his elbow height, because he uses it for writing purposes only. His feet should rest firmly on the ground, and his back should be supported firmly so that he can sit erect. Sounds familiar? Leto has guests every day to meet him. They should be sitting face to face with him at the table. For meeting and talking to people, direct eye contact is needed, as we have seen before. One should also ensure that Leto does not remain confined to his seat, as it's detrimental to his health. One way of doing this is to keep the filing cabinet that he needs to access frequently outside his reach! Sounds weird! But this will ensure that Leto gets up from his seat and walks, thus inducing some activity in his sedentary lifestyle. How to make the place interesting? We have a couple of options in hand. The furniture and seating layout could be changed every month. This will bring in some variety. Next, one could think of changing the theme on the walls. You can have a different set of photographs on the walls for different seasons. For summer, you can have flowers and people. For monsoon, you can depict different moods of the monsoon season. This variety in space can reduce monotony to a certain extent. It creates a framework for better experiences by the users.

5.3 Dichotomy in Space

Ms Kamal, a colleague of Leto, is in the mid-fifties. Everyone jokes with her because of her peculiar character. She wants to sit in the privacy of her office, but at the same time after working for some time in her office, she feels claustrophobic and rushes out of the office to talk to other people in the vicinity. She starts entering other people's offices to talk to them, saying that she is feeling isolated and bored (Figure 5.2). This dichotomous character of the common person in space is nothing unusual. We know that humans are social animals and cannot live in isolation. At times, they require their private space, but at the same time they also require interacting with people around and they get confused about what they want. It's from here that the concept of an "open office" system emerged, which helps in addressing this character of humans. You have your own cubicle, and while you sit you are in your private zone, with no contact with others. The moment you get bored, you just stand up, and you can see other colleagues and feel connected with them.

5.4 Colour and Music at Work

At the assembly line, Leto has observed that the job of packaging done by the staff is extremely repetitive and boring. They must do

Figure 5.2 Dichotomy in the workplace.

the same things throughout the day. Leto has noticed that after lunch their productivity declines, and many of them are unable to concentrate on their jobs. The result is that they make mistakes and there are rejections in the assembly line. Leto tried a "trick" one day. He brought his personal audio system and started playing some popular "instrumental" music very close to the assembly line, so that everyone could listen. Initially the staff were a little amused and it seemed that they were distracted, and their work suffered in terms of productivity. Interestingly, as the staff got used to the light music, they started enjoying their work. The music also isolated all the other sharp noises on the shop floor. In ergonomics, these types of slow and rhythmic music are called "wallpaper" music, as it covers up unwanted sounds just like the wallpaper covers up dents and markings on the wall. Apart from this, this music was also found to increase productivity and the number of rejects in the assembly line also decreased. The next trick Leto tried was the use of colour. The walls of the place were white but coated with dust, thus making them look ugly. So, in the visual field at the assembly line, what the staff see is nothing but a dirty white wall. Leto planned to convert the wall into a theme which is popular among the staff. After discussing with them, he found that everyone loves old regional movies of a particular type. So, if the props used in the movie like bicycles, lanterns, chariots, and swords are used to decorate the wall, along with the use of some colours which were prevalent in the movies, it would make the place interesting and not dull. It was done and the staff at the assembly line were extremely happy and cheerful; moreover, their productivity also increased. Many of them gave feedback that their sense of monotony, boredom or fatigue was also primarily reduced (Figure 5.3).

Figure 5.3 Use of music and colour to break monotony.

5.5 Body Clock and Efficiency at 4 pm

In the office, Leto observed one thing very peculiar. The working efficiency of all the staff is not the same throughout the day. In the morning after the office starts, the efficiency peaks at 10 am roughly. Everyone is glued to work, and no one is talking. This apparent focus is sporadic and is seen at 10 am, and again the maximum peak is seen at 4–4.30 pm just before leaving for home (Figure 5.4). The rest of the time, the efficiency is mixed, wherein people worked, took short breaks, talked to each other and so on. This sudden "arousal" among humans is very common at any workplace and is also guided by a biological clock in the brain, which regulates different physiological parameters of the body like heart rate, blood pressure, capacity to think, decision-making and so on. We must remember that we normal humans are all born to sleep at night and remain awake during the daytime. If you make people work beyond the normal waking time, it can be counterproductive for the common person. For people in specialised fields, they are trained to work under adverse conditions. It's possible through regular habits

Figure 5.4 Sudden arousal of staff before going home.

(like students do) to change the biological clock and remain awake during the night studying and sleeping during the daytime.

5.6 Lunchtime

Lunchtime at the office is a fun time for all. It's not only about food but also meeting people and talking to them. Here also some specific patterns are seen, in terms of group formation. Best friends prefer sitting face to face and in clusters. Thus, the seating arrangements need to be designed accordingly. There are introverts who prefer remaining alone and avoid being in any group. These types of people prefer sitting all alone and do not want eye-to-eye contact with anyone (Figure 5.5). The seating system should have arrangements for these types of people as well in the eating area. There are people who prefer some sort of entertainment like music, magic and dance. So, such arrangements should be made in the eating area if possible. These things recharge the staff so that they can start working with increased vigour in the second half. Leto has noticed that on those days where there are no performances or people are in a hurry to just finish their lunch due to extreme work pressure, staff find it difficult to work with increased vigour until the end of the day.

Figure 5.5 Seating pattern of people during lunchtime.

5.7 Claustrophobia in Elevator

Inside the elevator, the office staff feels suffocating. Many people suffer from this problem. Why does this happen and what is the remedy for the same? As humans, we are all comfortable in a closed space only when we can communicate with the outside world (Figure 5.6). That's why capsule elevators surrounded by glass panels are so popular and people have no problem with them. In the office elevator which is already enclosed, what is the solution? You cannot include glass panels. One way to correct this is to include a mirror or two. That would make the space look bigger. The person inside the elevator will also get a little distracted while looking at her/his image and as a natural tendency would try to groom her/himself. Another option is to paste 3D

Figure 5.6 Feeling of claustrophobia inside the elevator.

images which when one stares at the images, they seem to pop up. All these are an attempt to communicate with the person inside, which reduces the sense of claustrophobia.

5.8 The Torture Chamber of the Boss

Whenever Leto and his colleagues are called by the boss, there is a sense of tension which prevails. The employees can anticipate what's going to happen inside the office of the boss because of their prior experience. This "image" that is created has a name in ergonomics called *mental model.* Leto says that if you tell any employee that the boss is calling them, everyone immediately visualises similar things. You enter the office, the boss welcomes you and asks you to sit down. Once you take your seat, you are asked whether you would like tea or coffee. The next step would be the boss asking you what you have done in the past month. Until this point, it's all a similar picture, and then it varies from employee to employee. So, the grilling of the employees followed by subsequent scolding for not being able to meet targets is perceived by the employees as torture and the office as the torture chamber (Figure 5.7).

Figure 5.7 The torture chamber of the boss.

5.9 Recap

a. A boring workplace can be made interesting by adding variety to it.
b. People have dual behaviour in the workplace; they want privacy, and they want to socialise as well.
c. Colour and light music can make boring tasks more interesting.
d. The working pattern of people varies throughout the day, with efficiency increasing before going home.
e. A sense of suffocation in a closed space can be eliminated through communication.
f. People have images in the mind of spaces and experiences of interacting with people.

5.10 Practice Session

a. You have been to the supermarket and found some staff doing the job of arranging products on the shelves. They said that it's boring. What ergonomic solutions would you suggest?

Directions:

i. Introducing light music at the workplace.
ii. If possible, rotate the job with other jobs.
iii. Have some micro rests in between.

b. In your office, your boss is complaining that he feels isolated when he sits in his office. At the same time, he needs privacy for doing his work. Your ergonomic advice is solicited.

Directions:

i. Ask him to take microphones and come out of the office.
ii. He can have a glass cabin with curtains. The curtains can be removed at times to see what others are doing.
iii. He can listen to light music between work.

c. You must travel to the eighth floor of your office and take the elevator daily. It is closed and you are afraid to travel, as you feel suffocated. What ergonomic solutions can you suggest?

Directions:

i. Travel with someone and not alone.
ii. Mirrors can be used inside the elevator.
iii. Use some good pictures of nature which can be changed every week.
iv. Introduce a small jigsaw puzzle to engage people inside.

6
Back Home

6.1 Overview

This chapter provides the readers an overview of the ergonomic issues in the life of a common person while coming back home. The challenges, joy and different commitments in the life of the common person with its ergonomic connections are discussed. The issues of crowds on the streets in the evening, especially on Friday, and their ergonomic issues are discussed here. When the common person enters home, tired after the day's work, how ergonomics plays a role in her/his life and helps her/him freshen up is discussed. The same room, though it's confined, is where the common person finds peace and solace as it's his/her home. This is where he/she has his/her dinner and watches television, and the ergonomic elements in each of these are talked about.

6.2 Friday: The Weekend Joy

Today being the last day of the week and also the last day of the month in the office, everyone is charged up, including Viren. People had their salary credited to their accounts and everyone is happy to see their bank account "full". Viren received a message on his mobile phone in the evening after lunch that his salary has been credited to his account. The very nature of the message "your account has been credited by" does not always convey that salary has been credited. Many times, it could be some other payments for extra work that Viren does to supplement his earnings. It would be better if the message conveyed specifically the amount and the purpose for which the amount has been credited. Ergonomics suggests proper feedback be given to users in order to reduce chances of uncertainty and error in the long run. Viren

DOI: 10.1201/9781042009152-6

was excited and travelling, and he just heard the message on his mobile phone. When he tried to read the message, he could not, as he was travelling on the bus, which was vibrating because of some bumps on the road. The small font size of the text was not readable to Viren. Such context of use needs to be factored in while designing products for human use. One needs to know exactly how the user is going to use the product. Here the mobile phone manufacturer didn't think that Viren would need to read the text message inside a moving bus which would vibrate due to the road conditions.

In the office today, Viren and his colleagues were working very hard from the morning. The working efficiency of everyone was at the peak just before going home, which is normally seen in other days of the week also. Today that efficiency was greater. This type of behaviour is common in humans as their working efficiency does not remain constant throughout the day. The phenomenon is called "arousal" in humans and has implications in the manufacturing sector when it comes to enhanced productivity at the time of need or extreme production demands.

6.3 Crowd on Road and Control

When Viren came to Sadar at 7 pm with his friends, he found that the place was crowded. The crowd was mainly of young people of his age. They were all crowding at different eateries or roaming on the roadside. The vast majority of them were riding their motorbikes, and there was chaos on the main road. Pratik found one interesting thing, which is when people walk in a crowd, they walk in groups of two, three or four side by side. The crowd normally follows the route or the direction taken by others. He has also observed that every person in the crowd moves at a different pace. So, if he is behind the crowd and wants to move faster, he has often been slow, because the people in front do not always move fast. So, the faster you want to travel in a crowd, the slower you move.

At counters at the food cart and stalls, Pratik finds that people have a tendency of breaking the queue and moving closer to the person at the cash counter, so that they get their food fast. Interestingly, at a newly opened ice-cream parlour, they have established a nice way of controlling the crowd. The shape of the cash counter is such that the crowd is forced to stand in a queue. There is a curve, rather a concavity, in which only one person fits. The two parts (Figure 6.1) are convex. So, if people try to break the queue and stand at the convex end of the counter, they are far apart from the cashier and would not be able to reach him or draw his attention. So, they are forced to stand behind the person standing in the concave portion of the counter, and ultimately a queue is formed without the need for enforcing or telling the people to do the same!

Figure 6.1 Counter design to force users to queue.

6.4 Same Place but Different at Night

Viren travels to his office every day through Sadar, which is the main happening place in the town. In the morning, the same place looks like a sleeping beauty: calm, quiet, people organised and moving so diligently with no chaos. As it becomes dark, the same place turns into a different place than what he saw in the morning. Again, afternoons are different, with a few people from the local offices running around for lunch and light meals. That means the time of the day can change the characteristics of a place. Time is an important dimension which changes the character of a place. This is the reason why you are unable to recognise a railway station at midnight, even if you have travelled there before during the daytime. In our mind, the image of the place has been "etched" with respect to the time of the day that we saw it last. That is the reason why the same place looks different in the morning, afternoon and evening. In fact, many events we do not enjoy at any time of the day. We love watching an opera in the evening or at night. We might prefer a circus in the afternoon.

6.5 Along the Known Path and the Signages

Viren while coming back home from office explores the path that he travels. He often thinks that he has been travelling from home to the office along the same path for so many years, but still there are so many new things he has yet to explore. Today while coming back home from the office, Viren was focusing on the different types of signages on the road. He found that the signages and their positions are very important. All the arrows caught his attention today. Arrows pointing towards the right and left were easy to understand. When Viren carefully looked at the arrow pointing towards the road ahead, he became perplexed. The arrow pointing towards the road ahead, when read from the front of the board, was pointing towards "heaven". Viren had to stop and take some time to decode that this arrow pointing upwards is unlike the right and left arrows; it meant the road ahead. This is because we perceive everything in

3D perspective. We relate to what we do and see in our daily life, and any deviation makes us jittery. His next concern was that while the sun was setting behind the signage, he was unable to read the sign board. This is because of the eye trying to adapt to the bright sunlight and allowing less light to enter the eyes. A protective mechanism by which the amount of light entering the eye is reduced, thus making things a little dark right in front. This happens when you enter a movie hall from bright sunlight–you are unable to see anything in the dark.

6.6 Freshen Up and the Confined Space

Viren enters his house and opens the closet where he keeps his shoes and clothes. He goes to the bathroom to freshen up. Today he is a keen observer. He finds the bathroom very suffocating as he cannot see anything outside. But when he stands at the wash basin and faces the mirror, he feels a bit relaxed. He has never noticed this change before. When looking at himself in the mirror, he forgets about his confinement within the four walls of the bathroom. This is the reason why elevators have mirrors inside so that people do not feel suffocated. After freshening up, when Viren enters the living room, he feels much more relaxed because of his movement from the bathroom to the large room with lots of windows through which he can see the outside. Though it was dark, he could see the lights in the other houses and on the streets as well.

6.7 Different Clothes, Different Moods

Viren feels that the clothing that he wears has a bearing on his mood. If it's warm, then light colours keep him cool. If it's cold, then dark-coloured clothes tend to keep him a little warm. When he has met his sales target at his office and is rewarded by his boss, then he prefers wearing his favourite white-coloured shirt with black motifs on it. Viren wonders whether it's the clothes which dictate his mood, or the mood which dictates what type of clothing he should wear.

6.8 Dinner and the Television

Viren turned on the television news and sat at the dinner table with his family. He noticed that the television height is too low compared to his sitting height at the dining table. He was eating, talking to his family and at times staring at the news. He was amazed how he could do all three tasks of eating and enjoying the taste of the food, listening to the news and talking to his family members.

His son asked dad, "why can't we have a bigger television?" Viren has never thought of this before. He sat on the sofa and tried to see how big a television screen he could see. The current television seems really small, because if he moves his head slightly sideways, his eyes miss the screen. So, the screen should be as per his sitting height and distance from the sofa, and the breadth should be enough to accommodate horizontal movement of the head.

6.9 Recap

a. People expect feedback in a proper format when they are not in a conducive environment.
b. People form different groups while in the public domain depending on their preference for interactions.
c. Crowd flow is slowed down because everyone walks at their own pace.
d. Signages are information in space which needs to adhere to people's stereotypes.
e. Confined spaces can be made better by mirrors.
f. Clothing can improve one's mood.
g. People are capable of multitasking and parallel processing.

6.10 Practice Session

a. While going from home to the supermarket, list the simple ergonomic issues that you observe.

Directions:

i. Focus on the signage system along the way.
ii. List the specific changes in the place during the day and at night.

iii. Observe the crowd flow pattern.
iv. See how the queues at the shops can be organised without asking people to stand in a particular way.

b. List the ergonomic issues in your living room.

Directions:

i. Observe the furniture and its dimensions.
ii. See how the layout can be changed according to people's movement.
iii. Use the windows so that when you sit, you can see the outside.
iv. Try using indoor plants for breathing life into the space.

c. You sit for dinner every evening. Think of the ergonomic issues at the dining table and how you can improve upon the dining experience.

Directions:

i. Observe the height of chair and the table.
ii. Make sure there is enough clearance between people's elbows at the dining table.
iii. The food should be kept at a place where everyone should be able to reach it.
iv. Position the television so that everyone is able to see while dining.
v. Ensure people-to-people interaction.

7
Dusk

7.1 Overview

This chapter provides the readers an overview of the ergonomic principles involved when the common person prepares to go to bed. At the end of the day's work, the common person must retire to bed for a perfect sleep so that she/he is ready for the next day. It starts with the perfect-fitting nightdress as per the body movements and dimensions. Before retiring to bed, the person must plan for the next day in a manner that things move smoothly and she/he doesn't forget anything. The ergonomic principles in reminding people are talked about here. The room and its calmness are crucial for a good sleep, and the role of ergonomics in ensuring that is discussed. The habit of being engaged with digital devices like the phone and laptop and how it disturbs your sleep are discussed with specific ergonomic directions for addressing them. Other necessary items like water, medicine and an alarm clock and how they should be placed are crucial for spending a good night in bed and getting up the next morning. The dimension of the bed, with which we started in Chapter 1, is discussed and how the movement pattern of the body should be factored in while designing the bed. Lastly, as the common person falls asleep and enters the dream world, ergonomics as the subject of the "common person" keeps working, trying to make the busy life a little more humane!

7.2 The Nightdress for Sleeping

For sleeping, all of us prefer very loose clothing. Ms and Mr Bagga are a middle-aged couple. Both are working professionally and need a good sleep so that they are ready for the next day's work. After the couple has shifted to a new apartment, they

DOI: 10.1201/9781042009152-7

find that their bed has some problems. Mr Baggas' pajamas seem to be too long and tight. The result is that he is unable to sleep comfortably as the pajamas press against the waist and he feels numb. Similarly, the length is so long that when Mr Bagga wishes to go to the washroom at night, he has tripped and fell. This is a typical problem of ergonomics related to fitting the clothes to the body. It has a name, *anthropometry*, which is designing any product (here clothing) in accordance with the body dimensions. It also factors in the movements that the person would exhibit and keeps extra allowance for the same. When a person is sleeping, the body exhibits different types of movements. At times it curls, becomes straight, rolls on the side, takes a prone position, and spreads the arms and legs apart. The clothing should be able to accommodate all these movements. Thus, while designing the nightdress, one needs to factor in these movements and add extra allowance to the clothing. In ergonomics, clothing is believed to be the "second" skin of humans and covers and gives protection. The space between the skin and the clothing is known as the *microclimate*, and the fabric quality and other factors play an important role in making this microclimate comfortable for humans. Maintaining this climate is crucial for good sleep. Thus, the nightdress should be loose and a little short so that the body moves freely and so that someone like Mr Bagga does not fall while navigating to the washroom at night.

7.3 Planning for the Next Day

Ms Bagga gets anxious while she goes to bed as she starts thinking of the work to be done the next day, how it would be done; what if she forgets the first thing in the morning, etc. The common person also has this problem of getting too worried while going to bed, because he/she thinks of the next day and the things to be done. People forget because they are limited in their random access memory (RAM) capacity (also known as short-term memory). So that they do not forget, it's recommended that the things to be done the next day are organised. The best way to do this is to write down on a piece of paper what to do the next day and keep it next to the bed on the table having the alarm and water. Once Ms Bagga gets up in the morning, she can go through the list and then start the day. In this way, she need not worry about what she has forgotten or what is to be done the next day after she goes to bed. In fact, she can keep on adding to the to-do list throughout the day and then finally keep it near the bedside before going to bed. Another way to plan for the next day is to maintain a day-wise diary which can be used to add to the to-do list. In fact, these diary pages can act as a reference for what things were done before and how many of them need to be repeated once again. Proper planning helps with easy and better retrieval of data. This aspect of forgetting and retrieving what to do is like the books in the library. If you were asked to find the book you are looking for from a heap of books on a table, how would you feel? On the other hand, if the books are organised on the shelf in a specific order according to their accession numbers, it becomes extremely easy to retrieve them. If Ms Bagga does a little bit of planning, she will be able to sleep in peace and not spend sleepless nights thinking of what is to be done the next day!

7.4 Sleep and the Calmness of the Room

Good sleep is dependent upon many factors. One is that the person must be stress free before going to bed, and we have talked about how to organise and prepare for the next day. The room should be silent and preferably dark. For sleeping, a hormone called melatonin is needed. In bright light, the production of this hormone is decreased and sleeping becomes difficult or not possible. So having heavy and dark curtains in your bedroom is a must for proper sleep. There should not be any noise in the room. Too much noise distracts and prevents the common person from going to sleep. This is where the carpet in the room can help to a certain extent in preventing noise.

7.5 Switching Off Digital Screens

The light from all digital screens prevents good sleep. As stated before, it decreases the secretion of the hormone melatonin. So, make sure to switch off all digital gadgets at least two hours before going to bed. Reading books (print ones!) is a good habit before going to bed. The only downside to it is you need light, and that can interfere with your sleep hormone, though the extent of sleep disturbance would be minimal compared to the exposure to a digital screen. Ronni, the couple's only son, studying in standard eleven, is an avid book lover. He must read a few pages in bed every night before going to bed. He rarely complains about the problem of going to sleep because most nights he goes to sleep with the bedside light on and the book pages open on the bed! This is a habit a common person must inculcate. The habit is formed and you get addicted to it (good addiction), and then you cannot live without it. Ergonomics supports this type of addiction (rather good addiction) that not only helps you to learn something new but helps you learn it through a safe and better medium which also promotes better sleep.

7.6 Water and Medicine at Bedside, Reminder

When Roni goes to bed, he wakes up once or twice to go to the washroom. He navigates easily without switching on the washroom lights and then again comes back to his bed. In the meantime, he drinks a glass of water. This navigation, as we have seen in Chapter 1, is a "conditioning" that any common person is capable of after a few practice sessions. It's just like driving your car. When you start, you must look at the gear shift, but after a few practice sessions, you can navigate easily without the need to look at the gear shift. The couple needs to take a few medicines on an empty stomach in the morning. The bedside table should be within the reach of the couple from the bed while lying down. The best way to ensure that the couple can grasp a glass of water and the medicine is to calculate the distance while lying down on the bed and while sitting on the bed. Care must be taken to ensure that the bedside table and the elements on the table are far apart so that in the dark, while looking for the medicine, the water-filled tumbler doesn't fall and create a mess. Ideally all the elements should be reachable by the couple so that they are able to grasp them with their forearms folded at the elbow joint at ninety to one hundred ten degrees. This angle also ensures that they have proper control over the glass-filled tumbler and the medicine. The reminder paper or the diary should also be kept next to the water tumbler and medicine so that they are reminded of the things to be done the next day.

7.7 In Bed, Dark and My Position

Ronny while in bed sleeps in various positions. His favourite position is sleeping in a prone position, that is, sleeping on his belly. He even turns on the bed and changes the position of his body completely. Hence the design of the bed must account for such weird movements, else Ronny won't be able to move and might get up while sleeping. Again, the question of body dimension and the design of the bed comes to the forefront. Ergonomics again comes at the end of the day! A slight movement of the curtain was causing the moonlight to fall on Ronny's eyes. He tried to sleep but could not. Finally, he got up, pulled the curtain to cut off the moonlight and then tried sleeping . . . he was fast asleep and snoring!!

7.8 Dream Sweet Dream

Sleep and dreams are important in one's life. Ronny enters a deep sleep and starts dreaming "ergonomics". This is the most powerful of the applied sciences which ensures that the most evolved species on the earth, "the human", lives a life which is fruitful and satisfying. It keeps on working even when we are asleep. There are a large number of people who make products, space, communication and services all around us which works day and night even when we sleep. As they work, the applied science of ergonomics is all about people in action, always, never tired; they keep on working relentlessly. Whether it's the services of the ambulance carrying the sick, or it's the automated teller machine from where people need to withdraw money at midnight due to some emergency, it's all about ergonomics, which ensures how these services help humans achieve what they want, in the best possible manner and in a short span of time. People like you, me and Ronny all benefit from this applied science. The only thing is that we are not always aware of ergonomics' impact. As you develop your power of observation, you will see how ergonomics works all around you, from dawn to dusk.

7.9 Recap

a. Before going to sleep ensure that digital devices are switched off well in advance, as the body needs some preparation time to sleep.
b. Your bed dimensions are crucial for a good night's sleep.
c. All that you need in the early morning or at night should be within easy reach. Thus, your bedside table should be positioned accordingly.
d. Your bedroom should be quiet and dark to ensure proper sleep. Light interferes with sleep due to changes in certain sleep-inducing chemicals in the blood.
e. Ergonomics works even when we are sleeping.

7.10 Practice Session

a. List the ergonomic touch points while lying down on your bed.

   Directions:

   i. Look at your body and list which parts cover the bed.
   ii. Think of the ease of getting on the bed and getting down.
   iii. Consider that you move when you sleep.
   iv. If you use extra pillows, keep space for that.

b. You must visit a sick friend of yours who is in the hospital. He wants you to visit him early next morning. Using ergonomic principles, how would you prepare for the same at night?

Directions:

i. List what help you can render to your friend.
ii. Prioritise the list as per importance.
iii. Keep the list handy until you fall asleep.
iv. Make sure you keep it in the pocket of the shirt you are going to wear.

c. Your grandfather needs your help in setting up his room for a good sleep. From an ergonomics perspective, what would be your approach in helping him out?

Directions:

i. All the curtains should be pulled to make the room dark.
ii. Prepare his bed so that he is comfortable.
iii. Keep a glass of water close to the bedside.
iv. Remove all digital gadgets.
v. Keep a book or newspaper handy.

8

Activities with Ergonomic Directions

8.1 Overview

This chapter is dedicated to different activities with ergonomic directions. The variety of activities are aligned with the sequence of different chapters in the book. After going through all the chapters, the readers should start doing the activities. The ergonomic directions will help the readers in addressing the activities. After going through all the chapters and completing the activities, the readers should have a fairly good understanding of the science of ergonomics and how it could be applied in our daily life. Furthermore, this would evoke curiosity amongst the readers in taking it up further or encourage them to share it with others, thus promoting the science of ergonomics among the common persons for whom this subject has huge potential.

E1. You get up early in the morning only to find that the alarm bell is yet to ring. You get out of your bed and reach out for the glass of drinking water. As you are drinking the water, the alarm which was set to ring suddenly starts ringing and the glass tumbler falls from your hand, breaks into pieces and the water spills all over the floor. You manage to switch off the alarm, but your bedroom is already in a mess. From an ergonomic perspective, you must suggest what could have been done to prevent this "mishap" early in the morning.

Directions:

a. Your water should be stored in a wooden or metallic tumbler and not glass because your level of alertness is low when you get up in the morning.

DOI: 10.1201/9781042009152-8

b. The alarm clock should preferably not have a "shriek" sound but a pleasant tune, to which you would like to listen for some time before switching it off.
c. The tumbler should have a handle which you could hold onto, reducing the chances of the tumbler falling from the hand.
d. The alarm should have a coating of a fluorescent strip for locating it in the dark. The same could be done for the tumbler full of water for locating it in the dark.

E2. After having your morning cup of tea, you decide to go jogging in the park close to your house. Today you will be wearing your new pair of sports shoes which you purchased online and received yesterday. You put on your shoes and start walking towards the park. You enter the park and after running for two rounds find that you are unable to run any further as the shoes start getting too tight on your feet. From an ergonomic perspective, what could be the probable reason for this, and how might you address such issues in the future?

Directions:

a. While buying your shoes, you factor in the issue that your feet swell after walking for some time.
b. Sports shoes purchased for jogging purposes by common people should be a size bigger when purchased.
c. In case you face a problem like this as mentioned above, you have only one way to alleviate the swelling: lift your leg on a park bench, and the blood which pooled flows back and the swelling subsides.
d. While purchasing shoes, consider your feet length as well as the "depth" of the feet from the toes towards the heel region.

E3. While walking in the morning at the park which has just opened, you feel the urge to attend to the call of nature. You are desperately looking around for a washroom or any information related to the location of the washroom. You find a few people, and most of them are young couples. You are a little hesitant to ask these people and start wondering what the best solution is. From your understanding of the science of ergonomics, how would you address such a problem?

Directions:

a. While moving in a new space, people have certain "expectations" of elements like washrooms. Normally these are at the corners.
b. Try placing the washrooms as per people's expectations so that locating them becomes easier.
c. Information related to the washroom should "pop up" from the surroundings and should be preferably of a colour and form that contrasts with the surroundings.
d. Try using symbols for the washroom and not text. It is better if this is also given at the entrance to the park so that people know that the park has a washroom.

E4. Leto enters the washroom in the park and rushes to relieve himself. He finds the commode too high, and his feet start dangling while he sits. After he is done, he tries to look for the toilet paper roll and finds it in a different location. He is frustrated and gets up. He goes to the wash basin to wash his hands with warm water. As he turns the faucet, chilly water gushes out and he fiddles with the faucet to get warm water. From an ergonomic perspective, you must suggest ways and means of addressing these issues.

Directions:

a. The commode is like a chair while it is being used. The height of the same should be fixed, keeping in mind the shortest person.
b. Toilet paper is to be placed within easy reach of the user while seated on the commode. Here also the person with the shortest reach should be considered.

c. The faucet in the public domain is an element of frustration. It should have proper marking as to the direction for cold and hot water so that mixing of the two becomes easier.

E5. After Duggi has completed her morning walk in the park, she takes her car and goes to the nearby supermarket to get some groceries for the day. She goes to park her car but is unable to find an empty parking bay. She becomes a little annoyed and finally manages to enter the supermarket. It is crowded today, being a weekend, and she tries to navigate through the different sections of the store. She is unable to move with her trolley as there are other people with trolleys and people squatting to pick up stuff from the lowest rack in the store. She manages to complete her shopping and reaches the checkout counter, where she had to stand in a queue. Your advice is solicited as an ergonomist as to how the experience of Duggi can be made better.

Directions:

a. Information related to parking spaces should be available to users in advance, so that they can make the proper decision on time.
b. The design of store space should consider maximum crowd formation, their movements and areas of bottlenecks.
c. While moving between shelves with a trolley, one needs to consider the extra space to be taken by the trolley and the people who squat to find their products. Squatting leads to consumption of extra space if you visualise it from the top.
d. Deciding the height of the checkout counter should first consider the staff, as cashiers must work for long hours. Ideally it should be closer to the cashier's elbow height while sitting or standing so that they do not need to stoop forward, which leads to back pain.

E6. Viren is a bachelor and cooks his own food and makes his tea. Today Viren entered the kitchen to make a cup of coffee. The domestic help who comes to clean the house has misplaced the utensils. Viren is suffering from low back pain and finding it hard to cook. From an ergonomic perspective, what do you think are the ergonomic interventions possible in a kitchen in our homes?

Directions:

a. The kitchen platform should be at elbow height or a little below.
b. The chopping area, the cooking area and the washing area should ideally be in a sequence.
c. The depth of the platform should be such that it is easy to reach the farthest end without having to lean forward.
d. Different utensils and crockery can be categorised on different shelves to reduce their chances of being misplaced.
e. All kitchen items should be visible.

E7. A small party has been arranged today, and around fifty people have been invited. Mami and Nitai are trying to set up their living room to accommodate these people. The room contains three sofa sets, a centre table, a dining table, a few extra chairs, a television and a music system. From an ergonomic perspective, can you help this couple to arrange the place for the upcoming party?

Directions:

a. Try to map the flow of people in space and the amount of space they would occupy.
b. Locate any zones for bottlenecks, and if possible, remove any furniture in that place.
c. People-to-people interaction is facilitated by face-to-face seating arrangement. Try facilitating that.
d. Earmark the dining area from the sitting area, giving people the freedom to eat when they want. This would reduce unnecessary crowd formation in any specific area.

e. Advise the couple to reach out to every guest in their seats. Otherwise, they would crowd around to look for the host and create a bottleneck in the space.

E8. Battu is planning to travel to the city today from his village to appear for an examination. The examination centre is located in a building called "Peace", which is in the middle of the city Bhalobasha. He gets off the bus but is confused about which direction he should go. He finds everyone too busy to help him out. The only things he can see are some signs and symbols on the road. These signs and symbols make no sense to him as he is seeing them for the first time. What are the ergonomic issues which you think can help people like Battu navigate to his destination with ease?

Directions:

a. Any coded information like symbols and signs for public use needs to be designed looking at whether users can understand it with ease.
b. While moving in any space, try giving users a preview of where they are and then tell them what the space contains.
c. Symbols for amenities and facilities should be tested with people for proper understanding and should not have any ambiguity.
d. People are tense in the public domain when they travel, and hence they trust other people more than any information in space.

E9. The local market, which is there every Sunday, selling fruit and vegetables produced locally, has heard about you. They approach you to help in making the marketplace better for the customer in terms of increased sales of products. What ergonomic intervention would you recommend?

Directions:

a. Have separate sections for fruits and vegetables.
b. Arrange fruit and vegetable stalls as per the season.

c. Those products which have exclusive offers could be clustered together.
d. Include signage indicating which product is available in which area.
e. Number every stall.

E10. From sunrise to sunset, make an elaborate list of ergonomic principles that you come across unknowingly in your daily life. Indicate those places where ergonomic interventions are possible.

Directions:

a. List your touch points with assorted products and spaces.
b. Think of ergonomics in action in each of them along four lines, namely product, space, communication and services.
c. Explore how you can make each other better to make your life more productive and comfortable.
d. Think of people who are elderly and challenged and how they could be accommodated.

Bibliography

Bloswick, D. S., Villnave, T., & Joseph, B. (1998). Ergonomics. In *Sourcebook of occupational rehabilitation* (pp. 145–165). Boston, MA: Springer US.

Câmara, J. J. D., de Castro Engler, R., & de Oliveira Fonseca, P. (2010). Analysis and ergonomics of houses for elderly people. *Periodicum Biologorum*, *112*(1), 47–50.

Charness, N. (1998). Ergonomics and ageing: The role of interactions. *Gerontechnology*, *48*, 62–73.

de Carvalho, R. J. M. (2012). ERGOPOLIS: An ergonomics approach applied to a city. *Work*, *41*(Supplement 1), 6071–6078.

Deouskar, N. (2017). The impact of ergonomics on the productivity of people. *International Journal of Marketing & Financial Management*, *5*(6), 59–63.

Dianat, I., Molenbroek, J., & Castellucci, H. I. (2018). A review of the methodology and applications of anthropometry in ergonomics and product design. *Ergonomics*, *61*(12), 1696–1720.

Dul, J., Bruder, R., Buckle, P., Carayon, P., Falzon, P., Marras, W. S., & Van der Doelen, B. (2012). A strategy for human factors/ergonomics: Developing the discipline and profession. *Ergonomics*, *55*(4), 377–395.

Ferreras, A., García, C., Tortosa, L., Prat, J. M., Molla, F., Navarro, J., & Dejoz, R. (2007). Ergonomic adaptation of workplaces for people with disabilities: Methodology and examples. In *Challenges for assistive technology. AAATE 2007 conference proceedings* (pp. 136–140). Amsterdam: IOS Press.

Kohn, J. P., & Friend, M. A. (1993). Quality and ergonomics: The team approach to the occupational people factor. *Professional Safety*, *38*(5), 39.

Koirala, R., & Nepal, A. (2022). A literature review on ergonomics, ergonomics practices, and employee performance. *Quest Journal of Management and Social Sciences*, *4*(2), 273–288.

Murrell, K. (2012). *Ergonomics: Man in his working environment*. New York: Springer Science & Business Media.

Stanton, N., & Young, M. (1998). Is utility in the mind of the beholder? A study of ergonomics methods. *Applied Ergonomics*, *29*(1), 41–54.

Tosi, F., & Tosi, F. (2020). Ergonomics and design. *Design for Ergonomics*, *2*, 3–29.

Welford, A. T. (1987). Ergonomics of human relations. *Ergonomics*, *30*(1), 3–8.

White, C. M. (2008). Ergonomics: What is it? *The Bent of Tau Beta Pi*, *2008*, 24–27.

Zionchenko, V., & Munipov, V. (2005). Fundamentals of ergonomics. *Ergonomics: Major Writings*, *1*, 17–37.